THE DEVIL THEORY OF WAR

THE
DEVIL THEORY
OF WAR

An Inquiry into the Nature of History
and the Possibility of Keeping
Out of War

by

CHARLES A. BEARD

AUTHOR OF
PRESIDENT ROOSEVELT AND THE COMING OF THE WAR, 1941

CO-AUTHOR OF
THE RISE OF AMERICAN CIVILIZATION
AND OF
BASIC HISTORY OF THE UNITED STATES

GREENWOOD PRESS, PUBLISHERS
NEW YORK

PREFATORY NOTE

THIS volume (originally published by the Vanguard Press, Inc. in 1936) is a brief analysis of America's increasing involvement in World War I. Written by our father, Charles A. Beard, to indicate how the nation might, perhaps, avoid being drawn into another such contest, it appeared as a great new struggle was brewing in Europe, the ancient storm center.

A few years later, that storm broke. The attack on Pearl Harbor plunged the United States into World War II. Once peace was assured, and hitherto sealed files were opened for study by congressional investigators, Beard again analyzed the steps by which we had become involved in war—in *American Foreign Policy in the Making, 1932-1940* (1946) and in *President Roosevelt and the Coming of the War, 1941* (1948). Since then, the United States has entered upon a third and then a fourth costly struggle, both due to an involvement in Asia, bringing risks in both cases of further complication and escalation. This "entanglement," still continuing, would seem to give *The Devil Theory of War* a new timeliness. (See for this: Manfred Jonas, *Isolationism in America, 1935-1941*, Ithaca, N.Y., 1966, Chapter V, entitled "The Devil Theory of War.")

A brief prefatory note to the original edition said: "This volume is an expansion of three articles published in *The New Republic*, to which has been added supporting evidence in the form of documents." It was dated, "Springtime, 1936" and written in New Milford, Connecticut.

—WILLIAM BEARD and MRS. MIRIAM B. VAGTS
Scottsdale, Arizona; Sherman, Connecticut—1968

CONTENTS

Prefatory Note   v

I. More Light on "The Road to War."   11

II. War Is Our Own Work   17

III. The First Act in the War Drama   30

IV. The Second Act in the War Drama   40

V. The Third Act in the War Drama   55

VI. The Fourth Act in the War Drama
—The Plunge   89

VII. Learning From Bitter Experience   107

THE DEVIL THEORY OF WAR

CHAPTER ONE

MORE LIGHT ON "THE ROAD TO WAR"

PROLOGUE TO NEW HISTORY

THE revelations of the Nye munitions committee, especially in January and February, 1936, have given the American people a chance to know something real about themselves and their doings. These findings will mark a new epoch in the discussion of two questions: How did we get into the World War? How can we stay out of the next world war? And surely that is important for Americans, if anything is important.

These revelations are comparable in significance to the diplomatic revelations that came in 1917 and the following years, when the secret archives of Petrograd, Berlin and Vienna were torn open. Then, for the first time, the people of the world could know just what governments had been doing behind closed doors, in preparation for the great clash of arms.

In some respects, the Nye revelations are even more informing. They are not confined to diplomatic manipulations. They deal with economic persons as well as political persons. They deliver positive information showing the pressures exerted by economic interests upon the government of the United States which carried the nation into war. They present a Who's Who of the momentous years and show who did what. Respecting no other large phase of American history do we have such a meticulous record of persons and pressures involved in the shaping of policies, decisions and events.

Besides furnishing materials for an encyclopedic knowledge of the events under review, the Nye revelations ought to put on their everlasting guard all people who want to be intelligent. To Americans who lived through the years 1914-1918 they disclose the starkness of the ignorance that passed for knowledge and wisdom in those fateful days. They make the files of the newspapers that reported and commented on events look like the superficial scribblings of ten-year-old children, so far as the actualities behind the scenes were concerned. People who imagine that they "know what is going on" because they read great metropolitan dailies will discover by reading the Nye records the shallowness and ir-relevance of their "knowledge" gathered from the press in 1914-1917. In the course of forty or fifty years some of the lessons brought to light in the story of the march along "the road to war" may sink into the minds of even Respect-able Citizens.

NO HISTORY—OLD OR NEW—GIVES THE CERTAIN CAUSES OF AMERICAN INTERVENTION IN THE WORLD WAR

But a warning signal must be run up. The revelations of the Nye committee did not show who or what "caused" America's entrance into the World War. The establishment of that "cause" is a performance, in my opinion, utterly

13

beyond the powers of the human mind, at least given the state of our knowledge.

The patience of the reader must be invoked, for the problem of "causation" in human history seems to be even more perplexing than the questions of higher physics with which Millikan, Jeans, Eddington and Whitehead have wrestled.

It is impossible to conceive the cause of the war in terms of some person or persons and some action or actions standing alone—separated from other persons and actions. Physicists are now chary of assigning "causes" for this or that. For example, a chemist has ten chemicals in solution in a test tube. He adds an eleventh. A precipitation follows. Did the eleventh chemical "cause" the precipitation? Or was it "caused" by the reaction of the other ten to the event of addition? Why speak of cause at all in this connection? The chemist is likely to be content with mere description: Ten chemicals: the addition of the eleventh chemical: precipitation. It adds nothing to his *knowledge* to say that the eleventh chemical "caused" precipitation.

Human personalities and events are highly complicated. They have come out of a long historical past. They have immediate and remote relations to contemporaries. All the historian can do is to describe some of them in their more or less immediate relations. He may describe with considerable accuracy the conditions that made *possible* what happened. He can trace lines of relationships converging upon an event. He can show that some persons had more influence than others on decisions made, but he cannot measure exactly degrees of influence. He can reduce no total situation to an equation and make a Q.E.D. We cannot know who or what "caused" the War, but we can know about leading persons who took actions verging in the direction of war and about some of the actions that verged in the direction of war. We can be reasonably sure that some conditions and actions were decidedly "favorable" to that eventuality. That is not omniscience, but at the very least it is a great deal better than nothing.

For example, we can say on the basis of knowl-

15

edge that American bankers sought to induce
Woodrow Wilson to change his mind and lift
the ban on loans to Allied belligerents. We know
that through Lansing and McAdoo they warned
him of an economic crisis to come in case he refused. We know that he yielded and lifted the
ban. But did the bankers "cause" his action? Or
did the "cause" lie somewhere in the thought and
sentiments of Woodrow Wilson? Had he refused, what would have been the cause of the
refusal? Who feels competent to answer that
question? What would be the Q.E.D. of the
answer? We know some of the events. That
knowledge is a guide to future action. In the ab-
sence of omniscience, this is useful.

CHAPTER TWO

WAR IS OUR OWN WORK

EVERYBODY loves to have and use a scapegoat. For centuries man laid the evils of the world on woman. He did it with a vengeance. Now many women turn on man and make him the scapegoat. They say that man *is* war. Turnabout may be fair play, but the question of war is too complicated to be answered by retorts.

Shocking as it may seem to simple minds, any attempt whatever to "explain" war involves an interpretation of all human history since mankind began its career on this planet, and especially American history in 1914-1917. Hence we must inquire into the nature of history itself. To dodge it is to dodge the whole issue before us.

THE DEVIL THEORY OF WAR AND HISTORY

At the outset we must get rid of a false notion of human affairs.

Widespread is the conviction that wicked men make war—political or economic men or both. Under this conception, the masses of the people are viewed as loving peace. They want to go about their daily work, to earn a good living, to get along with their neighbors and to see everybody prosperous.

Cherishing good will, they pursue their callings, occupations, professions and interests, without thought of stirring up domestic or foreign troubles. If let alone in their pacific pursuits, war will never come. In every referendum on war in the abstract, they vote overwhelmingly against it, just as they vote against sin at church on Sunday.

But into this idyllic scene of the people engaged in "peaceful" pursuits, wicked politicians, perhaps shoved along by wicked bankers, burst with their war cries. They stir up the people.

They thrust arms into their hands and marshal them off to war. The politician, with the banker in the background, is a kind of *deus ex machina,* a strange kind of demon, coming from the nether region and making the people do things they would never think of doing otherwise. Or if the source of the trouble is not some wicked person, it is a wicked "force"—an impersonal "cause" of war. If this force could be eliminated or "cured" the cause of war would be removed, and the people would keep their Utopia in everlasting peace. All we need to do, then, is to drive the wicked from power or exorcise the evil spirit. This done, peace will reign.

Support for this simple view of peace and war comes from many economists. They seem to regard the politician as a kind of evil shaman that intervenes in economic affairs, without rhyme or reason or for selfish reasons of his own. Where he comes from, how he gets his power and why he interferes—these are questions seldom explored by theorists of this persuasion. Politicians interfere with the people in business. If they would stop, enterprise would boom along and everybody (almost) would be employed and happy. But this *deus ex machina* keeps breaking into the blissful domestic scene and upsetting the perfect balance of "the delicately adjusted economy."

Besides making domestic troubles, politicians willfully lead countries into war. And war is a senseless, uneconomic performance that destroys the balance of world economy and brings on panics. If politicians would stop interfering at home and stop making wars, the whole world would be busy, employed and (almost) happy.

Of course economists do not always put the case in these brief terms, but this is what their speculations amount to when boiled down to fundamentals. Politicians upset things at home, make wars abroad and disarrange the balance of production and exchange. Otherwise, peace and prosperity.

THE DEVIL THEORY WILL NOT HOLD WATER

Anybody who will stop singing this old tune

for a few minutes and examine it in the light of a few facts will see how childish it is. In the pursuit of peace, people are doing things that have a direct bearing on war. They are producing goods and offering them for sale. And such is the nature of machine industry under capitalist propulsion that there is usually a lot more goods to sell than the buyers can pay for. These are the so called "surpluses" of industry and agriculture. In some lines they may be small, in others, large. If these surpluses are not large at the moment, the potentials for increasing them are always present in machine industry. Given a chance to "make a profit," the owners of land and industries can usually manage to expand on short notice and pour out as many goods as the market will take.

Now it generally happens, even in the best of times, that the peaceful people, engaged in pursuing ways of peace, are looking hungrily for buyers—foreign as well as domestic. As a matter of fact, it is as patriotic to sell to a foreigner as to a fellow citizen. The pressure for opportunities to sell goods is an enormous peacetime pressure that comes out of the peaceful pursuit of peaceful pursuits.

The politicians and the bankers who are often accused of making wars do not come from some nether region, under this world of peaceful people so busily engaged in peaceful pursuits. They all come from among these very people. Apart from the chronic professionals, most politicians have engaged at one time or another in some peaceful pursuit. Often they are lawyers who have labored hard for business men and corporations devoted to turning out goods and making profits. Bankers are as a rule middlemen. Their function is to facilitate the operations of making and selling goods. They are in constant touch with such operations. They lend money to set them in motion. They smooth the ways of exchange, both domestic and foreign.

Neither politicians nor bankers operate in a vacuum. They do not intrude themselves upon the people from some magic world of their own. The politician seldom if ever conjures up any measure

or scheme of action from the vasty deeps of his
own mind. He works on suggestions from people
engaged in one or more pursuits of peace, or
on threats, pressures and orders from them.
Whether he keeps the peace or goes to war he is
acting under the stimulus of demands from
groups, classes and interests. His strokes of state
do not come out of an empty sky. He dwells in
no ivory tower. He reflects the ideas and wishes
of his constituents. The banker also lives right
down in the middle of things, amid the pushing
and shoving of the market-place. He, too, doesn't
play pinochle in an ivory tower. He watches for
chances to speed up the business of making goods
and selling them at a profit—this being the great
peacetime pursuit of the nation at large.

WAR AS AN OUTCOME OF PEACETIME PURSUITS

In the summer of 1914 the American people
were busy as ever making and trying to sell
goods at an advantage. But things were not going at top-notch speed. Business had slowed down
in the preceding winter, notwithstanding the

New Freedom. There was a great deal of unemployment in the cities. Manufactures were not moving swiftly enough to suit the makers. The prices of farm produce did not satisfy farmers.

In other words, Americans could do a lot more business than they were doing—peaceful business of making and selling goods, at an advantage. They had no idea, it is true, of going out frankly with a battle axe to drive bargains with other nations, but they were looking for customers. That was what they imagined themselves to be here for, at least principally. After all, what other purpose would more firmly engage the affections of a peaceful and practical people?

Then the big war broke with a bang. The American people got excited about it and evolved all kinds of passions, sentiments and theories pertaining thereunto. But in general they were peaceful and wanted to go on making and selling goods, at an advantage. They had been doing that before the War came, and to continue the performance seemed as natural as sunrise.

Very soon the Allies, for whom the seas were

open, began to buy steel, manufactures and farm produce rather heavily. Industrialists and farmers were pleased to sell. Workers were pleased to have jobs. Merchants and bankers were pleased to facilitate the transactions of purchase and payment. The Wilson administration was not unpleased to have business looking up. A congressional election was approaching and the sign of the full dinner pail would help keep Democrats in power. Keeping Democrats in power was deemed for the good of the country, by Democrats.

Of course there was a lot of confusion in the beginning. American short-term loans had to be paid in London. The stock market was bewildered. And the wise prognosticators in Wall Street could not forecast the future to their satisfaction. But after the first confusions were over, the peaceful business of making and selling goods spurted forward. It just so happened, perhaps unfortunately, that the best of new customers were the Allied governments engaged in an unpacific enterprise. But they were good customers.

They were in a hurry. They needed goods, in fact, very badly, and were not inclined to haggle too much over prices, commissions and fees.

At first also they could pay for their purchases. They had gold to send over. They could muster American, Canadian and other foreign securities, and sell them to raise cash. Despite the War, some British, French and Russian goods were imported into the United States, and these imports provided credits for paying export bills. But there was a limit to the gold, securities and imports available to pay for American goods and keep Americans busy at peacetime pursuits.

Apparently a pinch was felt very early in the rush, for the French government soon sounded out the National City Bank on the possibility of a loan or credit. That would help. It would make American money available to pay American business men and farmers, engaged in peacetime pursuits. Bigger sales, bigger profits, bigger wages, bigger prices and bigger prosperity. It looked good to everybody—manufacturers, farmers, bankers, wage-earners and politicians who

26

wanted to stay in office. Perhaps some tears were shed, but they were not as big as millstones and did not get in the way of making and selling at an advantage.

To those Americans who, for one reason or another, sympathized with the Entente Allies, the boom in business was doubly sweet. It piled up profits and commissions; it meant strength for the Allies' cause; it helped "to save civilization." But very few Americans engaged in the peaceful pursuit of making and selling had any thought at first of getting into the war. They were all for more and better business. It was nearly as simple as that, apart from the uproar of propagandists, sentimentalists and the intelligentsia.

The Democrats knew the country when in 1916 they flung up the slogan "He kept us out of war." The great majority of the American people doubtless wanted to continue their peaceful pursuit, making and selling goods.

It was simple and natural. But few realized how fateful in outcome their peaceful pursuit was to be. Few realized that war is not made by a *deus ex machina,* but comes out of ideas, interests and activities cherished and followed in the preceding months and years of peace. The notion that peace might make war did not enter busy heads.

The big question in 1915 was: How can the Allies "pay" for more and more goods, and enable Americans to follow peaceful pursuits happily?

New York bankers found the answer. They communicated it to Robert Lansing, Secretary of State, to William G. McAdoo, Secretary of the Treasury, to Colonel House, confidential adviser to the President of the United States. The question reached Woodrow Wilson. The answer reached him. Finding the solution agreeable, he approved it. The bankers' solution worked—for a time. Americans bought bonds to pay themselves for goods sold to the Allies. It was wonderful, the way it worked—for a time. Americans could keep on with their peaceful pursuits, with bigger and better prospects—for a time, for a time.

But in time the Allies were in another jam. They were in danger of losing the War or entering a stalemate that would defeat their ambitions. As Ambassador Page informed President Wilson in March, 1917, defeat for the Allies meant an economic smash for the United States. The following month, President Wilson called on Congress to declare war on the German Empire. So the United States entered the War and was at war.

War is not the work of a demon. It is our very own work, for which we prepare, wittingly or not, in ways of peace. But most of us sit blindfolded at the preparation.

CHAPTER THREE

THE FIRST ACT IN THE WAR DRAMA

THE STAGE AND THE ACTORS

IN THE summer of 1914 American industry and agriculture were all set to go. Business was not good. There were surpluses to sell. There were plants to expand for the production of more goods. New acres awaited the plow. Idle workers in cities searched for employment. A mild crisis in economy was already at hand. The opening of the war in Europe provided the great opportunity. New York bankers were eager to implement it.

Of course, very few Americans engaged in pursuing their peaceful callings had any idea of "what it was all about." Most of them were following their noses in their immediate routines. That was the "natural" order of things.

But a few economic leaders near the center through which goods flowed to the Allies and payments flowed in exchange had a broader and longer view—not very broad, not very long, but still more extensive than that of "the unfeathered bipeds" around them. They were in a position to accelerate and direct the movement of commodities and payments, out of which prices, wages and profits eventuated. Acceleration was their business, and acceleration was to the immediate advantage of their fellow participants.

At the big center in Washington were the political leaders who had power over the policies of the United States government. Among them were President Wilson, Secretary McAdoo of the Treasury, Secretary Bryan of the State of Department, Robert Lansing, Counselor to the Department, Colonel House, private channel of communication for and with the President, and members of the directorate of the new Federal Reserve Banks. To all appearances they could throw the weight of the government of the United States one way or another, unless forsooth they were automata moving under the propulsions of others. At all events, there they were. The stage was set.

Now, thanks to the revelations of the Nye committee, the dark velvet curtain of history is raised on this stage and these actors. We can know today words and deeds that were unknown to more than a handful in those fateful years. The play is divided into acts. It looks like a Greek tragedy. It seems to flow to its deadly dénouement, yet with moments of apparent pause and indecision when a little turn this way or that might have effected a different outcome. Who knows?

ACT I. AUGUST 10, 1914–OCTOBER 22, 1914. THE GOVERNMENT OF THE UNITED STATES BLOCKS POWERFUL INTERESTS EAGER TO MAKE MONEY FOR THEMSELVES AND OTHERS.

On August 10, 1914, W. J. Bryan, Secretary of State, informed President Wilson that the

Morgan Company, of New York, had asked whether there would be any objection to their making a loan to the French government and to the Rothschilds (perhaps for the French government). Mr. Bryan expressed to the President the opinion that "money is the worst of all contrabands," that if loans were made to belligerents the passions of pecuniary interest would be enlisted in the war, and that "powerful financial interests" would be thrown into the balance, making it difficult to maintain neutrality. In other words, Mr. Bryan clearly foresaw and portrayed the consequences of lending money to belligerents.

President Wilson evidently agreed with him. On August 15, 1914, Secretary Bryan wrote to the Morgan Company saying that inquiry had been made as to the attitude of his government. This attitude he then summarized: "Loans made by American bankers to any foreign nation which is at war are inconsistent with the true spirit of neutrality."

DOCUMENT 1

SECRETARY BRYAN'S LETTERS BARRING "LOANS"
TO BELLIGERENTS, AUGUST, 1914

Here are two letters by William Jennings Bryan, Secretary of State, brought forth by an inquiry from the Morgan Company relative to money-lending. They established the principle that loans to belligerents were contrary to public policy. The first letter was dated August 10, 1914. Cross-heads in italics are inserted by the author.

My dear Mr. President:

I beg to communicate to you an important matter which has come before the department. Morgan Company of New York have asked whether there would be any objection to their making a loan to the French government and also the Rothschilds—I suppose that is intended for the French government.

I have conferred with Mr. Lansing and he knows of no legal objection to financing this loan, but I have suggested to him the advisability of presenting to you an aspect of the case which is

34

not legal but I believe to be consistent with our attitude in international matters. It is whether it would be advisable for this government to take the position that it will not approve of any loan to a belligerent nation. The reasons that I would give in support of this proposition are:

Money Is the Worst Contraband of War

First: Money is the worst of all contrabands because it commands everything else. The question of making loans contraband by international agreement has been discussed, but no action has been taken. I know of nothing that would do more to prevent war than an international agreement that neutral nations would not loan to belligerents.

While such an agreement would be of great advantage, could we not by our example hasten the reaching of such an agreement? We are the one great nation which is not involved and our refusal to loan to any belligerent would naturally tend to hasten a conclusion of the war. We are responsible for the use of our influence through example and, as we cannot tell what we can do until we try, the only way of testing our influence is to set the example and observe its effect. This

35

is the fundamental reason in support of the suggestion submitted.

Loans to Belligerents Will Sharpen
Domestic Divisions

Second: There is a special and local reason, it seems to me, why this course would be advisable. Mr. Lansing observed in the discussion of the subject that a loan would be taken by those in sympathy with the country in whose behalf the loan was negotiated. If we approved of a loan to France we could not, of course, object to a loan to Great Britain, Germany, Russia, Austria or to any other country, and if loans were made to those countries our citizens would be divided into groups, each group loaning money to the country which it favors and this money could not be furnished without expressions of sympathy.

. These expressions of sympathy are disturbing enough when they do not rest upon pecuniary interests—they would be still more disturbing if each group was pecuniarily interested in the success of the nation to whom its members had loaned money.

36

*Loans Will Place Powerful Financial Interests
on the Side of Debtors*

Third: The powerful financial interests which
would be connected with these loans would be
tempted to use their influence through the newspapers to support the interests of the government
to which they had loaned because the value of the
security would be directly affected by the result
of the war. We would thus find our newspapers
violently arrayed on one side or the other, each
paper supporting a financial group and pecuniary
interest. All of this influence would make it all
the more difficult for us to maintain neutrality, as
our action on various questions that would arise
would affect one side or the other and powerful
financial interests would be thrown into the
balance.

I am to talk over the telephone with Mr.
Davison of the Morgan Company at 1 o'clock,
but I will have him delay final action until you
have time to consider this question.

It grieves me to be compelled to intrude any
question upon you at this time, but I am sure you
will pardon me for submitting a matter of such
great importance.

37

With assurance of high respect, I am, my dear
Mr. President,

Yours very truly,

W. J. BRYAN.

P. S. Mr. Lansing calls attention to the fact
that an American citizen who goes abroad and
voluntarily enlists in the army of a belligerent
nation loses the protection of his citizenship while
so engaged, and asks why dollars, going abroad
and enlisting in war, should be more protected.
As we cannot prevent American citizens going
abroad at their own risk, so we cannot prevent
dollars going abroad at the risk of the owners,
but the influence of the government is used to
prevent American citizens from doing this.
Would the government not be justified in using
its influence against the enlistment of the nation's
dollars in a foreign war? The Morgans say that
the money would be spent here, but the floating
of these loans would absorb the loanable funds
and might affect our ability to borrow.

DOCUMENT 2

Having consulted President Wilson and received his views, Secretary Bryan placed a

38

positive ban on "loans" to belligerents in the following letter to the J. P. Morgan Company:

Department of State,
Washington.

Aug. 15, 1914.

J. P. Morgan & Co.,
New York City,

Inquiry having been made as to the attitude of this government in case American bankers are asked to make loans to foreign governments during the war in Europe, the following announcement is made:

"There is no reason why loans should not be made to the governments of neutral nations, but, in the judgment of this government, loans by American bankers to any foreign nation which is at war is inconsistent with the true spirit of neutrality."

(Signed) W. J. BRYAN.

CHAPTER FOUR

THE SECOND ACT IN THE WAR DRAMA

BLOCKED temporarily by Secretary Bryan's letter, powerful banking interests besought the government of the United States for some kind of ruling that would lift the Bryan ban, at least to some extent. If they could not facilitate business for themselves and others by loans, they might do it by "credits." To obtain government consent to "credits" became their immediate concern.

ACT II. OCTOBER 23, 1914–MIDSUMMER, 1915. BANKERS PUT THE GOVERNMENT OF THE UNITED STATES IN A DILEMMA AND WIN APPROVAL FOR CREDITS TO BELLIGERENTS

On October 23, 1914, a vice-president of the

National City Bank of New York (apparently Samuel McRoberts) conversed with Robert Lansing of the State Department and sent him a letter. He told Mr. Lansing that his bank wanted to stimulate the unprecedented and increasing volume of business which came out of buying by foreign governments and their nationals. American manufacturers who were customers of the bank were urging the institution to provide temporary credits for foreign buyers of their goods. If American credits were dried up, the business would go to other countries. "If we allow these purchases to go elsewhere we will have neglected our foreign trade at the time of our greatest need and greatest opportunity. . . . The business which I have attempted to describe to you we deem necessary to the general good." So much for the vice-president of the National City Bank.

The first crisis was at hand. No bankers' credits for belligerents, a slowing down of their buying. A drop in buying, a loss of business. A loss of business, falling demand, prices, wages, employment and profits, assuming that nothing else could be done.

A key person at the center acts promptly.

That very night at 8:30, Robert Lansing called on President Wilson. Apparently he did not take up this vital issue with his chief, Secretary Bryan, but went behind the back of the State Department's head. Anyway, he saw President Wilson that night and discussed with him a difference between "loans" and "arrangements for easy exchange in meeting debts incurred in trade between governments and American merchants." The matter of "an arrangement of this sort [for credits] ought not to be submitted to this government." But President Wilson is willing for the bankers to go ahead and extend credits. Such were the "impressions" gathered by Mr. Lansing in his conversation with President Wilson.

To be sure, the question of "such an arrangement" had been submitted to the government of the United States. That was the subject of Mr. Lansing's conversation with the President. And the President "authorized" Mr. Lansing to give these "impressions" to such persons as were "entitled to hear them, upon the express understanding that they were my own [Mr. Lansing's] impressions and that I had no authority to speak for the President or the government."

To say the least, that was strange business. The question of credits to belligerent governments had been submitted to the Counselor of the State Department, but it ought not to have been submitted. President Wilson considered it, gave his opinion on it, "authorized" Mr. Lansing to transmit the impressions as his own to persons "entitled to hear them," on the understanding that they were Mr. Lansing's impressions, given without ascribing them to the President or the government. They were fateful impressions, and the American people were evidently not "entitled" to hear them.

At 8:30 P.M., on October 24, 1914, Mr. Lansing conveyed "the substance" of the "impressions" to a gentleman at the Metropolitan Club. This gentleman held some kind of place in the

House of J. P. Morgan, but Mr. Morgan was not quite sure about its nature in 1936.

At 10:30 A.M., on October 26, 1914, Mr. Lansing conveyed the substance of his "impressions" to an agent of the National City Bank, at the State Department. But Frank Vanderlip, of the National City Bank at the time, could not remember anything definite about it in January, 1936.

Anyway, the deed was done. Bankers could go ahead full steam providing credits for foreign buyers of American goods. And they did, thus accelerating the tempo of American industry, agriculture and merchandising — to the immediate advantage of all parties, so it seemed.

DOCUMENT 3

BANKERS BRING PRESSURE TO BEAR ON THE
GOVERNMENT OF THE UNITED STATES TO
AUTHORIZE CREDITS TO BELLIGERENTS

Just what Secretary Bryan meant when he put a ban on "loans" to belligerents was not explained in his letter to J. P. Morgan. Did he include short-term bank loans, short-term credits, and other forms of money lending? Or did he mean merely loans in the form of bonds sold to the American public for the benefit of belligerents?

Whatever may be the answer to these questions, one thing is certain. About two months after Secretary Bryan put the ban on "loans," a representative of the National City Bank in New York got in touch with Robert Lansing, Counselor of the State Department, and brought to his attention the pressing grounds for the extension of credits to the Entente Allies. The representative was apparently Mr. Samuel McRoberts, vice-president of the Bank. Why did he not take up the matter with Secretary Bryan? It involved a policy which the Secretary had been instrumental in establishing. There is no answer, at least at the moment. Besides talking with Mr. Lansing, Mr. McRoberts wrote him the following letter in which he showed that the sellers of American goods would derive advantages from the extension of credits and would suffer disadvantages if the ban was placed on credits. It was a case of "favor our request or the country will lose money." Cross-heads in italics are inserted by the author.

National City Bank,
55 Wall St.,
New York City, N. Y.

Oct. 23, 1914.

To the Hon. Robert Lansing,
Counselor, State Department,
.Washington, D. C.

Mr. Counselor:

Supplementing our conversation of this morning, I desire to call your particular attention to the following conditions now existing in this country and abroad.

The outbreak of the European war came at a time when this country owed a large amount to Europe, particularly to England, in the form of short time drafts, maturing between the outbreak of the war and the end of the year. The amount, while large, was not abnormal, considering the volume of our trade relations and was directly due to the anticipated shipment of cotton during the Autumn.

.War conditions, as you are aware, have made

46

cotton bills unavailable for the settlement of this balance against us and it can only be wiped out by the shipment of the goods, in lieu of the cotton, that are now needed and desired by the various European countries.

This is true, regardless of any temporary bridging over of the situation, and it has been the policy of the National City Bank, as far as possible and proper, to stimulate the unprecedented and unusual buying that is now going on in this country by foreign governments and their nationals. Since the beginning of the war this bank alone has received cabled instructions for the payment of in excess of $50,000,000 for American goods and the volume of this business is increasing.

Belligerents in Need of Credits

Owing to war conditions this buying is necessarily for cash, and it is of such magnitude that the cash credits of the European governments are being fast depleted. Lately we have been urged by manufacturers who are customers of the bank, and in some cases by representatives of the foreign governments, to provide temporary credits for these purchases.

47

For that purpose we have recently arranged to advance the Norwegian Government some $3,-000,000, practically all of which is to be expended for cereals in this country. Very recently the Russian Government has placed directly, and through agents, large orders with American manufacturers—such large orders that their cash credit has been absorbed and they have asked us to allow an overdraft, secured by gold deposited in their State bank, of some $5,000,000.

Some of our clients have been asked to take short-time Treasury warrants of the French Government in payment for goods and have, in turn, asked us if we could discount them, or purchase warrants direct from the French Government for the purpose of replenishing their cash balances. We have also been asked by European interests practically the same question as to English consols and Treasury securities. Some of our German correspondents have approached us with the suggestion that, without naming a particular security, we sell securities to increase their cash account with us, and we have little doubt this is indirectly for the purpose of the German Government.

48

Credits or Loss of Trade

We strongly feel the necessity of aiding the situation by temporary credits of this sort, otherwise the buying power of these foreign purchasers will dry up and the business will go to Australia, Canada, Argentina and elsewhere. It may in the end come back to us, but the critical time for American finance in our international relations is during the next three or four months, and if we allow these purchases to go elsewhere we will have neglected our foreign trade at the time of our greatest need and greatest opportunity.

It is the desire of the National City Bank to be absolutely in accord with the policies of our own government, both in its legal position and in the spirit of its operations, and, while very anxious to stimulate our foreign trade, we do not wish to, in any respect, act otherwise than in complete accord with the policy of our government.

For the purpose of enabling them to make cash payments for American goods, the bank is disposed to grant short-time banking credits to European governments, both belligerent and neutral, and, where necessary or desirable, replenish their cash balance on this side by the pur-

49

chase of short-time Treasury warrants. Such purchases would necessarily be limited to the legal
capacity of the bank, and as these warrants are
bearer warrants without interest, they could not
and would not be made the subject of a public
issue. These securities could be sold abroad or be
readily available as collateral in our foreign loans
and would be paid at maturity in dollars or equivalent in foreign exchange.

Business for the General Good

This business which I have attempted to describe to you we deem necessary to the general
good, and we desire to proceed along the lines in-
dicated unless it is objectionable from the government's standpoint, in which case we assume that
you will advise us.

Very respectfully yours,

. . . .

Vice President.

DOCUMENT 4

COUNSELOR LANSING GOES TO THE WHITE
HOUSE WITH THE BANKERS' PRESSING QUESTION.
CREDITS EXTENDED TO BELLIGERENTS

50

On the very night of the request from the agent of the National City Bank, Counselor Lansing called on President Wilson at 8:30. He had a talk with President Wilson. Just what the line was, nobody knows. On his return to the State Department at 9:30 that night, Mr. Lansing wrote the following memorandum of his conversation with the President. Thus a distinction was made between Allied bond issues floated in the United States and "an arrangement for easy exchange in meeting debts" incurred in sales to belligerents. President Wilson did not authorize Mr. Lansing to speak for him, did not go on record as approving anything. He merely allowed Mr. Lansing to convey his own impressions of what might have been in the President's mind— convey them "to such persons as were entitled to hear them." At the bottom of his memorandum, Mr. Lansing added a note to the effect that he had given these "impressions" to a representative of the J. P. Morgan Company and to a representative of the National City Bank. These representatives, apparently, were "entitled to hear them."

Lansing Memorandum

The State Department memorandum was as follows:

Department of State,
Office of the Counselor,
9:30 P. M.

Oct. 23, 1914.

Memorandum of a
Conversation with Pres.
At 8:30 this evening
Relative to loans and
Bank credits to belligerent governments.

From my conversation with the President I gathered the following impressions as to his views concerning bank credits of belligerent governments in contradistinction to a public loan floated in this country.

There is a decided difference between an issue of government bonds, which are sold in open market to investors, and an arrangement for easy exchange in meeting debts incurred in trade between a government and American merchants.

The sale of bonds draws gold from the American people. The purchasers of bonds are loaning their savings to the belligerent government, and are, in fact, financing the war.

The acceptance of Treasury notes or other evidences of debt in payment for articles purchased in this country is merely a means of facilitating trade by a system of credits which will avoid the clumsy and impractical method of cash payments. As trade with belligerents is legitimate and proper it is desirable that obstacles, such as interference with an arrangement of credits or easy method of exchange, should be removed.

The question of an arrangement of this sort ought not to be submitted to this government for its opinion, since it has given its views on loans in general, although an arrangement as to credits has to do with a commercial debt rather than with a loan of money.

The above are my individual impressions of the conversation with the President, who authorized me to give them to such persons as were entitled to hear them, upon the express understanding that they were my own impressions and that I had no authority to speak for the President or the government.

ROBERT LANSING.

53

Substance of above conveyed to Willard Straight at Metropolitan Club, 8:30 P.M., Oct. 24, 1914. Substance of above conveyed to R. L. Farnham, at the department, 10:30 A.M., Oct. 26, 1914.

CHAPTER FIVE

THE THIRD ACT IN THE WAR DRAMA

By the midsummer of 1915 the Allies had apparently stretched their short-term "credits" to the limit. At all events the British pound was slipping down. The British government and the Morgan Company either could not or would not sustain it. If it was not sustained a smash was in sight—for busy Americans engaged in producing manufactures and farm produce, and for other parties.

ACT III. AUGUST 14, 1915–AUGUST 26, 1915. BANKERS PUT THE GOVERNMENT OF THE UNITED STATES IN ANOTHER DILEMMA AND OBTAIN APPROVAL FOR LOANS TO BELLIGERENTS

August 14, 1915, Benjamin Strong, of the New

York Federal Reserve Bank, wrote to Colonel House, called his attention to the slipping sterling pound, and remarked that "the influence is gradually growing stronger to curtail our export business."

August 17, 1915, J. B. Forgan, president of the First National Bank of Chicago, wrote to F. A. Delano, vice-governor of the Federal Reserve Board, saying that he wanted to get some information for "a very confidential purpose." The Allies needed more than credits; they needed a loan. What would be the government's attitude on a loan?

Copies of Mr. Forgan's letter reached Secretary McAdoo, of the Treasury, and Secretary Lansing, now head of the State Department in place of W. J. Bryan.

August 21, 1915, Mr. McAdoo wrote a long letter to President Wilson on the gravity of the economic situation. He told the President bluntly that: "Our prosperity is dependent upon our continued and enlarged foreign trade. To preserve that we must do everything we can to assist our customers to buy. . . . To maintain our prosperity we must finance it. Otherwise we must stop, and that would be disastrous." The upshot: the Bryan ban on loans stood in the way of prosperity; indeed, threatened the country and the Wilson administration with a disaster.

August 23, 1915, Secretary McAdoo sent a copy of J. B. Forgan's letter to Secretary Lansing, called his attention to the points raised by it, condemned by implication the Bryan ban on loans, and said that he would see him in person in a few days.

August 25, 1915, Secretary Lansing wrote to President Wilson, enclosing a copy of Mr. Forgan's letter, expressed the opinion that changed conditions must be recognized, and declared that "the large debts which result from purchases by belligerent governments require some method of funding these debts in this country."

Another hour of fateful decision had come. The President of the United States was informed by bankers and by his official advisers that a domestic crash would come if the huge debts created by the credits he had obliquely approved in October, 1914, were not funded. In other words, unless American investors now put up the money to pay Americans for American goods bought by the Allies, the outcome would be "disastrous." Having taken the step to credits in October, 1914, to prevent a slowing down of business, he must now approve funding these credits and others into American loans to avoid a disastrous slowing down—an economic calamity.

Confronting this dilemma, President Wilson chose a course. Perhaps there was no choice. Perhaps given his outlook on life, on human values and on national destiny, the constrictions of iron pressures determined the matter in his mind. Who knows?

However that may be, on August 26, 1915, President Wilson conveyed to Secretary Lansing his decision in another letter also oblique but sufficient: "My opinion in this matter, compendiously stated, is that we should say that 'parties would take no action either for or against such a transaction,' but that this should be orally conveyed, so far as we are concerned, and not put in writing." A copy of this letter Secretary Lansing confidentially transmitted to Secretary McAdoo.

Soon the news was "conveyed orally" to persons "entitled" to hear it. Bankers had their release authorization to go ahead full steam, facilitating the transactions that kept the American people busy and happy in their peaceful pursuits. Investors among them, including manufacturers profiting from the business, put up millions and millions to "pay" the American people for American goods.

DOCUMENT 5

BENJAMIN STRONG, GOVERNOR OF THE NEW YORK FEDERAL RESERVE BANK, WARNS COLONEL HOUSE THAT MORE MONEY IS NEEDED BY THE ALLIES

By mid-summer, 1915, the bankers were in trouble again. Belligerent buyers were finding difficulty in obtaining sufficient credits to sustain their heavy buying. The British sterling pound was slipping because the British government and the Morgan Company would not or could not sustain it. A sharp decline in the pound would curtail buying in the United States and cut down "our" export business. That would make domestic economic trouble. August 14, 1915, Benjamin Strong, governor of the New York Federal Reserve Bank, conveyed this information to Colonel House who was a private channel of communication with the President of the United States. Here is the warning letter:

Aug. 14, 1915.

My dear Colonel House:

Referring to our conversation of a week ago; you have doubtless observed that matters are developing along the lines of our discussion. Sterling exchange sold yesterday below 4.71. The newspapers are reporting very considerable cancellations of foreign contracts for grain, which is reported to be due to military developments at the Dardanelles, which may shortly release large quantities of Russian wheat.

This seems hardly probable, and, if rumors now appearing in the newspapers are well-grounded (although I suppose they are considerably exaggerated), I am inclined to believe that the cause is inability to get remittances. It is a striking illustration of the possible effect upon our trade growing out of inability to arrange credits in this country.

If exchange declines very sharply so that all the profit on a purchase of goods contracted for in this country is gone before the goods are exported and the purchaser is in a position to cancel the contract, he will, of course, cancel in every instance even though he has to buy again later, possibly after contracting for his exchange in advance.

The situation is undoubtedly growing increasingly difficult with each day's decline in exchange and, while I don't see anything yet to be alarmed about, I still believe that at present rates, with the prospect of still lower rates, the influence is gradually growing stronger to curtail our export business.

With kindest regards, I beg to remain,

Very truly yours,

BENJAMIN STRONG, JR.

DOCUMENT 6

A CHICAGO BANKER PUTS THE ISSUE OF LOANS
TO BELLIGERENTS UP TO THE GOVERNMENT OF
THE UNITED STATES

By the middle of August, 1915, it was evident
that the Allies had stretched their "credits." They
must have a loan or they could not go on buying.
That meant that American investors must buy
bonds from the Allies and thus furnish the money
to pay Americans for goods bought. On August
17, 1915, James B. Forgan, president of the Chicago First National Bank, wrote to F. A. Delano,
vice-governor of the Federal Reserve Board, laid
the question of loans before him, and asked him
to find out what position the Government of the
United States would take. Here is the letter:

Chicago, Aug. 17, 1915.
My dear Mr. Delano:

I want to get some information for a very confidential purpose and it has occurred to me that
you may be in a position to help me secure it.

It is, to put it bluntly: I would like to know
what the attitude of the government administra-

62

tion in Washington would be toward the flotation of a large British loan in this country. Some time ago I remember seeing in the press that the State Department had discouraged New York bankers on a proposition to float a British loan in this country, but at the same time it was stated that it was not within the province of the government to veto such a transaction.

It would seem to me that the present condition of international exchange would deter the government from entering any objection to the flotation of such a loan in this country, or to the sale by Great Britain of American securities in this country. One or the other of these transactions would seem to be a business necessity at the present time.

As I am in a bit of a hurry to get the information, I would appreciate a telegram indicating what you believe the government's attitude would be. You might send me one of the following telegrams to indicate which of the positions you think the government would take in regard to the flotation of a large British loan in this country and I will understand your meaning:

1. Parties would be favorable to and would encourage such a transaction.

2. Parties would take no action either for or against such a transaction.

3. Parties would discourage such a transaction but would not offer any active interference with it.

4. Parties' attitude would be such as to make such a transaction practically impossible.

With kind regards, I am,

Very truly yours,

JAMES B. FORGAN.

DOCUMENT 7

W. G. MCADOO, SECRETARY OF THE TREASURY, PUTS THE ISSUE OF LOANS UP TO THE PRESIDENT OF THE UNITED STATES

In a long, but blunt, letter, dated August 21, 1915, the Secretary of the Treasury, W. G. McAdoo, informed President Wilson in effect that prosperity would burst if the Allies could not get American money through loans to pay Americans for goods bought. The Bryan ban must be broken down and prosperity saved. Mr. McAdoo's letter runs as follows (with cross-heads in italics, inserted by the author):

Aug. 21, 1915.

Dear Governor:

You know how loath I am always to burden you with Treasury affairs, but matters of such great importance have arisen in connection with the financing of our export trade that you ought to know the facts.

Great Britain is, and always has been, our best customer. Since the war began, her purchases and those of her allies, France, Russia and Italy, have enormously increased. Food products constitute the greater part of these purchases, but war munitions, which, as you know, embrace not only arms and ammunition, but saddles, horses and mules and a variety of things, are a big item.

War Makes Prosperity

The high prices for food products have brought great prosperity to our farmers, while the purchases of war munitions have stimulated industry and have set factories going to full capacity throughout the great manufacturing districts, while the reduction of imports and their actual cessation in some cases, have caused new industries to spring up and others to be enlarged.

Great prosperity is coming. It is, in large measure, here already. It will be tremendously increased if we can extend reasonable credits to our customers. The balance of trade is so largely in our favor and will grow even larger if trade continues that we cannot demand payments in gold alone, without eventually exhausting the gold reserves of our best customers which would ruin their credit and stop their trade with us.

They must begin to cut their purchases from us to the lowest limit, unless we extend to them reasonable credits. Our prosperity is dependent on our continued and enlarged foreign trade. To preserve that we must do everything we can to assist our customers to buy.

Munitions Trade Is Lawful Commerce

We have repeatedly declared that it is lawful for our citizens to manufacture and sell to belligerents munitions of war. It is lawful commerce and being lawful is entitled to the same treatment at the hands of our bankers, in financing it, as any other part of our lawful commerce.

Acceptances based upon such exportations of goods are quite as properly the subject of legitimate bank transactions as if based on non-contraband. We have reaffirmed our position about munitions in our recent note to Austria, clearly and conclusively.

If our national banks are permitted to purchase such acceptances freely, it will greatly relieve the situation. They can do so without any danger of rendering "non-liquid" even a small part of our present extraordinarily large credit resources. But national banks will not buy such acceptances freely unless they know that they are eligible for rediscount at Federal Reserve Banks.

Federal Reserve Board Has Blocked Munition Financing

We have two strong pro-German members of the Reserve Board—Messrs. Miller and Warburg. Miller is even stronger pro-German than Warburg. He is a far less reasonable and intelligent man, and more difficult to deal with. He and Warburg have always been opposed to our banks buying acceptances or accepting them for the exportation of munitions, although they know that it is perfectly lawful for the banks (members as well as Federal Reserve) to do so, and counsel for the board has so advised. The board has no power whatever under the Federal Reserve Act to discriminate between lawful ex-

67

THE DEVIL THEORY OF WAR

ports or imports or the acceptances based thereon.

While I was ill and confined to my bed, the board (Federal Reserve) on April 2, 1915, adopted "Regulation J" on bankers' acceptances. Warburg and Miller skillfully injected into these regulations such restrictions and limitations as to, in effect, render ineligible for rediscount in Federal Reserve Banks a large part of the acceptances based on munitions exports.

I never saw these regulations, as I was too ill at that time to consider business, and the other members of the board seem to have acted without full realization of the effect of the regulations. These regulations (I mean Sub-sections C and F) are clearly ultra vires; they should never have been adopted.

Five of us favor repeal or modification, but it brings the issue sharply and concretely to the front. Messrs. Miller and Warburg are insisting that the administration define its position or attitude on the question of financing war munitions. It is not necessary for them to do this, i. e., raise such an issue. If they were thinking of our interests instead of Germany's, they would not do it. I believe that their purpose is to embarrass, if possible, the administration and that it is deliberate.

It ought to be sufficient that the administration has declared exports of munitions lawful commerce and that the Federal Reserve Act clearly makes acceptances based thereon lawful business for national banks and Federal Reserve Banks—leaving the policy of buying or dealing in or creating such acceptances to the officers and directors of the banks—national and Federal Reserve —to determine.

I think that Mr. Miller and Mr. Warburg both agree to the soundness of this as a legal proposition—although they have succeeded in imposing the limitation on the Federal Reserve Banks to rediscount the acceptances in question by a skillfully disguised definition of "bankers' acceptances"—Sub-sections C and F of the regulations of April 2, 1915.

These Sub-sections will have to be rescinded. Governor Hamlin writes me that both Miller and Warburg have intimated plainly to him "that they would both oppose such rescission." I understand that Warburg has gone so far as to say "that financing of war munitions, although legal in form, violates the spirit of neutrality."

The exchange situation in New York had become so serious that it became necessary to call a hurried meeting of the Federal Reserve Board in New York on Aug. 10 to clear up the question of acceptances by national banks, which, although a matter for the Controller of the Currency to decide, it was important for the Reserve Board to concur in so that national banks could proceed with the knowledge that acceptances so made or bought by them would be eligible for rediscount by Federal Reserve Banks.

War Financing Hampered by Bryan's Ban Against Loans

Without knowing that Governor Hamlin had authorized a call for a meeting at the Federal Reserve Bank, I directed that the meeting be called for the Sub-Treasury, my idea being that as the questions upon which the board was asked to rule were presented by the Federal Reserve Bank of New York and it was interested in the result, it would be better for the board, which is a government body with its headquarters in the Treasury at Washington, to meet in the United States Sub-Treasury in New York instead of in the bank itself, which was seeking a decision of the board.

I could not remain for the meeting. As you know, I had to see House promptly about some very urgent matters.

Miller, Warburg and Harding refused to attend at the Sub-Treasury. Hamlin and Williams had to join them at the bank to make a quorum.

I attach a report of Mr. Williams on this meeting which is very truthful and dispassionate, and I wish you would read it. Hamlin also sent a report, but Williams's is, on the whole, better because more in detail.

I also attach a letter from Governor Harding. His quotation from Warburg's letter to him is most interesting. Warburg doesn't want England to buy cotton; prefers a "valorization scheme" rather than put ourselves at "England's mercy." He seems to favor our buying up the "war materials" produced in this country. Harding's reply demolished him.

Warburg also complains because Hamlin and Delano conferred with Colonel House, and thinks he ought to be fully informed, but he doesn't offer to inform the board of the conversations with Bernstorff and Albert (the chief figure in the

71

World exposures), with each of whom he is very intimate—especially with Albert. I hope you may have time, some time, to read Harding's letters to me and Warburg.

Governor Hamlin writes me: "Mr. Miller is strongly opposed to any acceptances being issued by national banks or rediscounted by Federal Reserve Banks involving the exportation of war materials. He puts it rather on the ground of policy than any prohibition in the law." But to get back to the meeting: The Controller and the Federal Reserve Board agreed on a satisfactory solution of the acceptance problem for member banks, but the question is yet to be dealt with for Federal Reserve Banks.

I shall insist on the repeal of the restrictive regulations of April 2, 1915, and I shall refuse to say in response to Miller and Warburg that the administration expressly desires Federal Reserve Banks to finance exports of war munitions; that it is solely for the directors and officers of the Reserve Banks to determine that matter for themselves. This being done, our export trade will get substantial relief but not nearly enough.

It is imperative for England to establish a large credit in this country. She will need at least

72

$500,000,000. She can't get this in any way, at the moment, that seems feasible, except by sale of short-time government notes. Here she encounters the obstacles presented by Mr. Bryan's letter of Jan. 20, 1915, to Senator Stone, in which it is stated that "war loans in this country were disapproved because inconsistent with the spirit of neutrality," &c., and "this government has not been advised that any general loans have been made by foreign governments in this country since the *President expressed his wish that loans of this character should not be made.*"

The underscored (italicized) part is the hardest hurdle of the entire letter. Large banking houses here which have the ability to finance a large loan will not do so or even attempt to do, in the face of this declaration. We have tied our hands so that we cannot help ourselves or help our best customers. France and Russia are in the same boat. Each, especially France, needs a large credit here.

The declaration seems to me most illogical and inconsistent. We approve and encourage sales of supplies to England and others, but we disapprove the creation for them of credit balances here to finance their lawful and welcome pur-

73

chases. We must find some way to give them needed credits, but there is no way, I fear, unless this declaration can be modified. Maybe the Arabic incident may clarify the situation. I should hate to have to have it modified that way.

Notwithstanding Mr. Bryan's letter expressing disapproval of foreign loans, the German Government openly issued and sold last Spring through Chandler Brothers, bankers of Philadelphia and New York, $10,000,000 of its short-time bonds. England and her allies could sell a small amount of obligations, perhaps $25,000,-000, in the face of your disapproval as expressed in this letter, but it would be fruitless. The problem is so huge that she must go "whole hog," and she can't do that unless our attitude can be modified.

"Prosperity" or Disaster

Perhaps it could be done, if you decided that it should be done at all, by some hint to bankers, although I do not think that would do. In fact, England and her allies will have great difficulty in getting the amount of credit they need here even if our government is openly friendly. I wish you would think about this so we may discuss it when I see you. To maintain our prosperity, we must finance it. Otherwise it may stop and that would be disastrous.

I have not the slightest fear that we shall be embarrassed if we extend huge credits to foreign governments to enable them to buy our products. Our credit resources are simply marvelous now. They are easily five to six billion dollars. We could utilize one billion in financing our foreign trade without inconvenience and with benefit to the country.

I wrote Lansing a brief note yesterday about credits to foreign governments and suggested that nothing be done to emphasize the position taken in Mr. Byran's note until I could have a chance to discuss it with you and him.

My apology for writing at such length is my desire to give a clear idea, if possible, of the problem and the situation in the Reserve Board which has within it mischievous potentialities on account of the attitude of Miller and Warburg.

Later on I may submit to you my correspondence with them about their refusal to attend a lawfully called meeting of the board in New York. Thus far they have evaded the issue and I am without a direct statement from them of the reasons for their extraordinary and inexcusable action.

Affectionately yours,

W. G. McADOO.

DOCUMENT 8

SECRETARY MCADOO BRINGS THE CHICAGO
BANKER'S QUESTION TO THE ATTENTION OF
SECRETARY LANSING

By some process Mr. McAdoo received a copy of Mr. Forgan's letter (above, p. 62) and two days after he had written to President Wilson he sent a copy of the Forgan letter to the Secretary of State (now Mr. Lansing) with the following communication:

Aug. 23, 1915.

Confidential.

Dear Mr. Secretary:

I enclose copy of a letter from James B. Forgan of Chicago to Vice Governor Delano of the Federal Reserve Board in reference to the matter of foreign loans in this country. The foreign exchange situation is so serious that it may become imperative for some of the foreign governments to establish credits in this country in order that they may continue to purchase freely our farm products and other supplies.

The attitude of the government, as expressed in the letter of Secretary Bryan to Senator Stone, Jan. 20, 1915, may seriously embarrass the creation of such credits in favor of foreign governments as are needed to enable them to continue their purchases in this country. Germany, by the way, disregarded this letter and placed more than ten million of short-time notes in this country through Chandler Brothers of Philadelphia.

It is not my purpose, however, to discuss that; I only mean to direct your attention to the importance of giving very serious thought to the points raised in Mr. Forgan's letter. I have always felt that it was a mistake for our government to discountenance in any way the establishment of credits in this country in favor of foreign governments, such credits to be used in purchasing supplies in this country.

It seems to me entirely inconsistent to say that the purchase of our farm products and manufactured articles and other supplies by foreign governments is lawful and to be encouraged, and then to say that we discourage and discounte-

77

nance as being unneutral the credit operations which are an essential part of such transactions.

I merely desire to call your attention at the moment to the seriousness of the question and to say that I hope no action will be taken that will add to the embarrassment of the situation by reaffirming or emphasizing the position taken in Mr. Bryan's letter of Jan. 20, last, until I have had an opportunity to discuss this with you and the President.

I shall certainly be in Washington on the first of September—maybe sooner. I look forward with pleasure to seeing you then.

With warmest regards, I am,

Faithfully yours,

WILLIAM G. McADOO.

DOCUMENT 9

ROBERT LANSING, SECRETARY OF STATE, PUTS THE
NECESSITY OF LOANS TO BELLIGERENTS UP TO
PRESIDENT WILSON

Four days after Secretary McAdoo gave President Wilson the choice between loans and economic crisis, Secretary Lansing, who had succeeded Secretary Bryan, sent the President a copy of Banker Forgan's letter and asked for advice. He did not present alternatives to the President. He declared flatly that the state of affairs required some method of funding the large debts already incurred by belligerents:

Aug. 25, 1915.

My dear Mr. President:

As the letter of Mr. James B. Forgan, which is enclosed to me by Mr. Hamlin, deals directly with the general policy of the government I feel that before answering it I should be advised as to your wishes. I therefore enclose Mr. Hamlin's letter and a copy of Mr. Forgan's.

I think we must recognize the fact that conditions have materially changed since last Autumn, when we endeavored to discourage the flotation of any general loan by a belligerent in this country. The question of exchange and the large debts which result from purchases by belligerent governments require some method of funding these debts in this country.

Faithfully yours,
ROBERT LANSING.

79

DOCUMENT 10

PRESIDENT WILSON LIFTS THE BAN ON LOANS
TO BELLIGERENTS

In response to the requests made by Secretary
McAdoo and Secretary Lansing, President Wilson made the fateful decision to permit the flotation of loans for belligerents. His decision was
transmitted to Secretary Lansing in the following letter incorporating Banker Forgan's own
formulation:

Aug. 26, 1915.

My dear Mr. Secretary:

My opinion in this matter, compendiously
stated, is that we should say that "parties would
take no action either for or against such a transaction," but that this should be orally conveyed,
so far as we are concerned, and not put in writing. I hope that this is also your own judgment
in the matter.

Faithfully yours,
(Initialed) W. W.

DOCUMENT 11

SECRETARY LANSING INFORMS SECRETARY
MCADOO OF PRESIDENT WILSON'S DECISION IN
FAVOR OF LOANS TO BELLIGERENTS

On the day that President Wilson's letter was received, Secretary Lansing wrote the following note to Secretary McAdoo.

Aug. 26, 1915.

Confidential.

My dear Mr. Secretary:

Mr. Hamlin sent me a copy of the letter of Mr. James B. Forgan which you enclosed to me in your letter of the twenty-third. I sent the letter to the President on the twenty-fifth; a copy of my letter to him is enclosed, and I also enclose his reply of today.

I have read your comments upon the matter of loans to belligerent countries and must say that I concur in your opinion—in fact, from the outset I have held that opinion of such transactions, viewed from the legal standpoint rather than from the standpoint of expediency.

While the President did not authorize me to.

send a copy of his communication to you I feel that he would wish you to know his position.

Faithfully yours,

ROBERT LANSING.

DOCUMENT 12

SECRETARY LANSING AND PRESIDENT WILSON
REINFORCE THE POLICY OF ALLOWING LOANS TO
BELLIGERENTS

As if the decision on Banker Forgan's letter had been insufficient to formalize and clinch the policy of permitting the flotation of belligerent loans in America, Secretary Lansing sent a longer letter on the subject to President Wilson and received a confirming reply. This exchange of correspondence follows (with cross-heads in italics inserted by the author):

Lansing's Letter to Wilson

Mr. Lansing's letter to President Wilson was as follows:

Sept. 6, 1915.
My dear Mr. President:
Doubtless Secretary McAdoo has discussed

82

with you the necessity of floating government loans for the belligerent nations which are purchasing such great quantities of goods in this country, in order to avoid a serious financial situation which will not only affect them but this country as well.

Briefly, the situation as I understand it is this: Since Dec. 1, 1914, to June 30, 1915, our exports have exceeded our imports by nearly a billion dollars, and it is estimated that the excess will be from July 1 to Dec. 31, 1915, a billion and three-quarters. Thus for the year 1915 the excess will be approximately two and a half billions of dollars.

It is estimated that the European banks have about three and one-half billions of dollars in gold in their vaults. To withdraw any considerable amount would disastrously affect the credit of the European nations, and the consequence would be a general state of bankruptcy.

If the European countries cannot find means to pay for the excess of goods sold to them over those purchased from them they will have to stop buying, and our present export trade will shrink proportionately. The result would be restriction of outputs, industrial depression, idle capital and idle labor, numerous failures, financial demoralization and general unrest and suffering among the laboring classes.

Trouble Ahead

Probably a billion and three-quarters of the excess of European purchases can be taken care of by the sale of American securities held in Europe and by the transfer of trade balances of Oriental countries, but that will leave three-quarters of a billion to be met in some other way. Furthermore, even if that is arranged, we will have to face a more serious situation in January, 1916, as the American securities held abroad will have been exhausted.

I believe that Secretary McAdoo is convinced, and I agree with him, that there is only one means of avoiding this situation which would so seriously affect economic conditions in this country, and that is the flotation of large bond issues by the belligerent governments.

Our financial institutions have the money to loan and wish to do so. On account of the great balance of trade in our favor, the proceeds of these loans would be expended here. The result would be a maintenance of the credit of the borrowing nations based on their gold reserve, a continuance of our commerce at its present volume and industrial activity, with the consequent employment of capital and labor and national prosperity.

The difficulty is—and this is what Secretary McAdoo came to see me about—that the government early in the war announced that it considered "war loans" to be contrary to "the true spirit of neutrality." A declaration to this effect was given to the press about Aug. 15, 1914, by Secretary Bryan. The language is as follows: "In the judgment of this government, loans by American bankers to any foreign nation at war is inconsistent with the true spirit of neutrality."

The Past Reviewed

In October, 1914, after a conference with you, I gave my "impressions" to certain New York bankers in reference to "credit loans," but the general statement remained unaffected. In drafting the letter of Jan. 20, 1915, to Senator Stone I sought to leave out a broad statement and to explain merely the reasons for distinguishing between "general loans" and "credit loans." However, Mr. Bryan thought it well to repeat the

August declaration and it appears in the first sentence of Division 13 of the letter, a copy of which I enclose.

On March 31, 1915, another press statement was given out from the department, which reads as follows:

"The State Department has from time to time received information directly or indirectly to the effect that belligerent nations had arranged with banks in the United States for credits in various sums. While loans to belligerents have been disapproved, this government has not felt that it was justified in interposing objection to the credit arrangements which have been brought to its attention. It has neither approved these nor disapproved—it has simply taken no action in the premises and expressed no opinion."

Manifestly, the government has committed itself to the policy of discouraging general loans to belligerent governments. The practical reasons for the policy at the time we adopted it were sound, but basing it on the ground that loans are "inconsistent with the true spirit of neutrality" is now a source of embarrassment.

This latter ground is as strong today as it was a year ago, while the practical reasons for discouraging loans have largely disappeared. We have more money than we can use. Popular sympathy has become crystallized in favor of one or another of the belligerents to such an extent that the purchase of bonds would in no way increase the bitterness of partisanship or cause a possibly serious situation.

A Crisis at Hand

Now, on the other hand, we are face to face with what appears to be a critical economic situation, which can only be relieved apparently by the investment of American capital in foreign loans to be used in liquidating the enormous balance of trade in favor of the United States.

Can we afford to let a declaration as to our conception of "the true spirit of neutrality" made in the first days of the war stand in the way of our national interests, which seem to be seriously threatened?

If we cannot afford to do this, how are we to explain away the declaration and maintain a semblance of consistency?

My opinion is that we ought to allow the loans to be made for our own good, and I have been seeking some means of harmonizing our policy,

87

so unconditionally announced, with the flotation of general loans. As yet I have found no solution to the problem.

Secretary McAdoo considers that the situation is becoming acute and that something should be done at once to avoid the disastrous results which will follow a continuance of the present policy.

Faithfully yours,

ROBERT LANSING.

President Wilson's Reply

Sept. 8, 1915.

My dear Mr. Secretary:

I have no doubt that our oral discussion of this matter suffices. If it does not, will you let me know that you would like a written reply?

Faithfully yours, .W. .W.

CHAPTER SIX

THE FOURTH ACT IN THE WAR DRAMA—THE PLUNGE

HAVING broken down the ban on credits and then on loans, bankers made the most of their opportunity. Loan after loan was floated to pay Americans for American goods. As the days and weeks passed the fate of American bankers, manufacturers, farmers, merchants, workers, and white-collar servants became more deeply entangled in the fate of the Allies on the battlefield—in the war.

ACT IV. SEPTEMBER, 1915–APRIL, 1917. DEEP IN THE MIRE OF WAR THROUGH LOANS, CREDITS, AND SALES, THE GOVERNMENT OF THE UNITED STATES ENTERS THE WAR

By the opening of 1917 the Allies had stretched

89

their borrowing power—perhaps not to the limit. Who knew the limit? But it was stretched. On the field of battle things were not going well. In March, 1917, the Tsar's government collapsed, threatening to take Russia out of the War. Defeat was possible for the Allies or a stalemate equivalent to a defeat of their ambitions expressed in the Secret Treaties dividing the spoils of victory in advance of victory.

Again, the Wilson administration faced a crisis such as it had faced in October, 1914 and again in August, 1915. If the war stopped, American business would slow down from prosperity to dullness, if not calamity. If the Allies were defeated, things would be worse. American millions were at stake. What other things were really at stake no one knew.

But a crisis was there—cold, brutal, remorseless. Economic leaders as well as political leaders, now all entangled in the same fateful web, were under a great strain. The propaganda for American participation in the War increased. Much could be said for it, and was said. Immediate advantages to stake-holders were apparent, surrounded by dark shadows of uncertainty.

Amid this tension, the German government renewed its submarine warfare. This action has been called "the primary and final cause" of the American declaration of war.

Why call it "the" cause? Why not make "the" cause the action of the Allied governments in imposing an "illegal" blockade on Germany? Germany alleged that this was the cause of her action called the cause of President Wilson's action. Why stop in the search for causes with the British action? Why not attribute the cause to the action of the United States government in acquiescing in British action? Why stop anywhere along the "chain of causes" short of the "first cause"? Nobody knows or will ever know "the" cause of Woodrow Wilson's decision in favor of war.

But we do know a great deal about the economic conditions and pressures that made his decision possible and the execution of that decision possible. The Nye revelations portray them in multitudinous detail and show powerful leaders in banking and politics taking actions that facilitated the slide toward the war abyss. In the letters and papers of these leaders written for private consumption the burden of all arguments was the maintenance of American prosperity and the fear of domestic calamity: Goods must be sold and paid for in some way or the swollen structure of prosperity would burst.

The President of the United States and his advisers knew this. They had weathered two crises by yielding to the demands of bankers for a free hand with credits and loans. Each time they had been told that it was a question of favorable action or an economic explosion at home. They knew in the Spring of 1917 that a still greater crisis of the same kind was at hand, and that there was choice between an economic explosion at home and making war abroad.

These were the circumstances in which the decision to enter the war was made. Walter Hines Page, American ambassador in Great Britain, set the circumstances clearly before President

Wilson in the following cable, dated March 5, 1917. The situation could not be more accurately described. As Mr. Page put the case, the J. P. Morgan Company seemed to be near the end of its power to keep the finances of the Allies going in the United States. The country faced an economic smash at home or intervention in war on the side of the debtors.

DOCUMENT 13

'AMBASSADOR PAGE STATES THE DILEMMA'

London, March 5, 1917—1 p. m.
(Received March 6, 3:20 a. m.)

The financial inquiries made here reveal an international condition most alarming to the American financial and industrial outlook. England is obliged to finance her allies as well as to meet her own war expenses. She has as yet been able to do these tasks out of her own resources. But in addition to these tasks she cannot continue her present large purchases in the United States without shipments of gold to pay for them, and she cannot maintain large shipments of gold for two reasons: First, both England and France

93

must retain most of the gold they have to keep their paper money at par; and, second, the submarine has made the shipping of gold too hazardous, even if they had it to ship. The almost immediate danger, therefore, is that Franco-American and Anglo-American exchange will be so disturbed that orders by all the allied governments will be reduced to the minimum, and there will be almost a cessation of trans-Atlantic trade. This will, of course, cause a panic in the United States. The world will be divided into two hemispheres, one of which has gold and commodities, and the other, which needs these commodities, will have no money to pay for them and practically no commodities of their own to exchange for them. The financial and commercial result will be almost as bad for one as for the other. This condition may soon come suddenly unless action is quickly taken to prevent it. France and England must have a large enough credit in the United States to prevent the collapse of world trade and of the whole European finance.

If we should go to war with Germany, the greatest help we could give the Allies would be such a credit. In that case our Government could,

94

if it would, make a large investment in a Franco-British loan or might guarantee such a loan. All the money would be kept in our own country, trade would be continued and enlarged till the war ends, and after war Europe would continue to buy food and would buy from us also an enormous supply of things to re-equip her peace industries. We should thus reap the profit of an uninterrupted, perhaps an enlarging, trade over a number of years, and we should hold their securities in payment.

But if we hold most of the money and Europe cannot pay for re-equipment, there may be a world-wide panic for an indefinite period.

Unless we go to war with Germany our Government, of course, cannot make such a direct grant of credit, but is there no way in which our Government might indirectly, immediately, help the establishment in the United States of a large Franco-British credit without a violation of armed neutrality? I am not sufficiently acquainted with our own reserve bank law to form an opinion, but if these banks were able to establish such a credit, they would avert this danger. It is a danger for us more real and imminent, I think, than the public on either side of the ocean re-

THE DEVIL THEORY OF WAR

alizes. If it be not averted before its symptoms become apparent, it will then be too late to avert it. I think that the pressure of this approaching crisis has gone beyond the ability of the Morgan Financial Agency for the British and French Governments. The need is becoming too great and urgent for any private agency to meet, for every such agency has to encounter jealousies of rivals and of sections.

Perhaps our going to war is the only way in which our present prominent trade position can be maintained and a panic averted. The submarine has added the last item to the danger of a financial world crash. During a period of uncertainty about our being drawn into the war, no more considerable credit can be privately placed in the United States and a collapse may come in the meantime.

PAGE.

Such were certain phases of American history immediately preceding the entrance of the United States into the war. Such were certain back-stage negotiations among heavily interested parties—bankers and politicians. Leaders in political affairs knew that a domestic crisis would flow, in all probability, from the defeat of the Allies or a stalemate that thwarted their ambitions. This is not to say that fear of the crisis was the "cause" of American intervention in the war. But indisputable evidence shows that in the autumn of 1914 and again in the summer of 1915 the issue of crisis or concession was put up to President Wilson by bankers and political advisers, and that President Wilson had yielded on both occasions—had chosen concession rather than crisis.

Evidence so far unearthed does not show that bankers put the issue of crisis or war up to President Wilson in the spring of 1917. What was said privately by bankers and advisers we do not know. They did not have to "see" him personally about this issue. Since President Wilson was a man of intelligence and knowledge, it is reasonably certain that he was then aware of the economic dilemma before him. If he had not developed that awareness out of his experience in the autumn of 1914 and the summer of 1915, then he could scarcely have escaped it on reading Am-

97

bassador Page's precise description of the dilemma in his message of March 5.

Yet we are told that "the" cause of American intervention was the renewal of the German submarine campaign. "There," says Charles Seymour, "lay the basic cause of American intervention. It is historically isolated, as one isolates a microbe." If there had been no German submarine campaign, there would have been no war.

Now it is possible to isolate a microbe, but it is not possible to isolate any event in history from other events, except in one's mind and theoretically. Events in history as reality are connected with other events, past and contemporaneous. The renewal of the German submarine campaign was one of the events that may have predisposed President Wilson in the direction of war. How "heavily" it "weighed" in his mind we do not know. Nor could President Wilson himself have known precisely.

One need only to try to imagine such an operation to discover the absurdity of the very idea. The idea is an analogy borrowed from physics—

a kind of animism. To carry it out we must conclude that a net weight of some kind tipped the "balance" in President Wilson's mind to war. Let us say that it was twelve ounces. If so, did the submarine event weigh two ounces, eight ounces, ten ounces, eleven ounces, twelve ounces?

Such an analogy is simply preposterous as a description of mental operations. Resort to astrology is quite as "reasonable."

All that we can say realistically is that the German submarine campaign was one of the events in a vast concatenation of events that shortly preceded American intervention in the war. It was one event in a total situation of events that made possible President Wilson's decision and action.

To travel beyond that is to express a mere opinion. To say that President Wilson could and would have taken the country into war on the submarine issue, if the posture of economic interests had not been favorable, is also to express an opinion—a dubious opinion. Indeed, why was the German submarine campaign followed by repercussions in the United States? In part, at least, for the reason that it smashed into the "profitable" business which American bankers and traders had built up with the Allies, and it threatened American bankers, traders, and other immediately and remotely interested parties with a crisis in domestic economy. In other words, the German submarine campaign was not an action from the blue sky which merely offended "American honor." It was part and parcel of the total military and economic situation. It was not isolated from the context of the times. The historian who isolates it in his mind is merely performing a mental trick that does not correspond to known realities.

Thus it is impossible to determine that the submarine campaign was "the" cause of American intervention. The case is not helped by adding question-begging adjectives, such as "basic," "immediate," "primary," or "fundamental." What is a "basic" cause, as distinguished from a cause? If such an adjective means anything, it means that the "basic" cause was not the whole or total cause. How does one discover the respec-

100

tive weights or values of the several parts of "the cause?" Let the proponents of the hypothesis make a demonstration. When they try to make answer they will discover the unreality of their figurative language.

This is not all. Historians who call the German submarine campaign "the basic cause" of American intervention must admit that they rely on evidence, on testimony of some kind. Well, on this point we have testimony of such high authority that few will dispute it. During its hearings, the Nye munitions committee introduced files of the Senate Foreign Relations Committee giving a colloquy between Senator McCumber and President Wilson shortly after the close of the war. It ran as follows:

SENATOR McCUMBER—Do you think that if Germany had committed no act of war or no act of injustice against our citizens we would have gotten into the war?

PRESIDENT WILSON—I think so.

SENATOR McCUMBER—You think that we

would have gotten in anyway?

PRESIDENT WILSON—I do.

What becomes of the German submarine campaign as "the basic or primary or fundamental cause" of American intervention? Such is the testimony of the President who went before Congress and asked for a declaration of war. Surely his testimony is to be taken as against the testimony of historians who imagine that they know the determining influence that operated in President Wilson's mind.

So we are back where we started. We do not know the "cause" of American intervention. We do know something about the operations of bankers and politicians which verged in the direction of war, which were favorable to war. We know that these operations were carried on secretly, that knowledge of them was not then revealed to the people of the United States or to the Congress of the United States. We know that these operations helped to entangle the fate of American economy in the fate of the Allied belligerents. We

102

T-32 8/87

BANK
Subsidiary of First Bancorporation of Ohio
Member FDIC

ITEMS DEPOSITED ARE SUBJECT TO
VERIFICATION AND CORRECTION

RECEIPT
PLEASE KEEP FOR YOUR RECORDS

Thank you for banking with us.

Trans. No.	Branch Bank	Teller	Mach.	Date	Time	Account No.	Transaction

0560*004 1001/14 12 NOV03-90 09:33 04150959 51.19 CK DP

24 HOUR
BANKER

know that bankers and politicians knew this, and had been active in creating the situation.

The question before us is not the question of their honor or wisdom or virtue. The question is simply this: Do we want, for the future, discussions and decisions of this character to be carried on secretly behind closed doors or openly in the Congress of the United States? In fine, are bans on loans, credits, and sales to belligerents to be raised clandestinely in huddled conferences of bankers and politicians or publicly by the representatives of the American people in Congress assembled?

That is the "lesson" of the last World War. That is the issue to be decided now.

THE BALANCE SHEET—DID IT PAY?

Did it "pay" in the end—this promotion of business enterprise—in the economic terms so dear to bankers, business men and all hardheaded persons? Did it "pay" the nation, as distinguished from profit-seeking bankers?

According to estimates of the Morgan Com-

103

pany, busy Americans sold the Allies about $7,000,000,000 worth of goods prior to April 1, 1917. It was this selling that produced the big "prosperity" which bankers and the Wilson administration tried so hard to uphold between 1914 and 1917.

It was seven billion dollars' worth of prosperity. Well, according to the estimates of President Coolidge, participation in the World War cost the people of the United States, besides death and suffering, at least $100,000,000,000, counting outlays to come for pensions, bonuses and other war charges.

It does not seem to have been in the long run a "paying" proposition, considered in the terms used by the bankers and politicians between October, 1914, and April, 1917. To be sure, J. P. Morgan was reported to have said in 1936 that "we saved our souls and civilization." But not much was said about souls and civilization in their private correspondence and papers during the period—August, 1914–January, 1917.

Just how much were the bankers, the politicians, the industrialists and the farmers in the play concerned with the salvation of souls? How much influence did the salvation of souls have on their weekday conduct?

DÉNOUEMENT

"We" went into the War and came out. As Al Smith would say: "Here we are"—in the middle of another economic crisis, and with the prospect of another world war and more of that kind of "prosperity" before us.

Have economic activities and interests changed fundamentally since 1914? Are we more concerned with the salvation of souls? Are our political and economic leaders men of the same type that was uppermost in 1914–1917? Do the American people really want to stay out of the next world war, to make the domestic changes necessary to that end? Or do they want to take another big gamble with "prosperity" and run the risk of a bigger and better burst?

A wit has said that the only thing we learn from history is that we learn nothing from it.

But at least it may be said that from the Nye papers we can know something about the history of 1914-1917 and the eminent gentlemen who played such a large part in making it. And the American people did not have to wait a hundred years to obtain this knowledge. For the opportunity thus vouchsafed, others besides dust-sifting historians ought to be grateful.

Low lights, soft music and curtain.

CHAPTER SEVEN

LEARNING FROM BITTER EXPERIENCE

HAVING before us the scenes described by the Nye committee, what should be the policy of the United States government in respect of the future? In those scenes we see powerful economic and political personalities seeking to avoid one domestic crisis after another by extending credits and loans to the Allies. At last in the presence of the third and major crisis the President of the United States avoided an immediate domestic crash by leading the country into war. Thus the smash was postponed for about ten years. It took place in 1929.

Similar conditions are still present. American industry, agriculture and banking "need business." They need it badly. For years they have been feverishly searching for more foreign markets. We are in the presence of a domestic distress far greater than that of 1914. Similar economic pressures are here; similar leaders are here, eager and willing to channel them into profitable transactions. The munitions business and allied enterprises are humming and have facilities for more output. The American people, at least those who have jobs, are following the pursuits of peace as in 1914-1917. European powers are heaping their armaments mountain high and once more tremble on the verge of a war. These appear to be facts in the present scene.

Should the United States government follow the policy pursued in 1914-1917—permit and encourage selling freely to belligerents, extending credits to them, making loans to them, and to neutrals engaged in selling to belligerents? Or should some new policy be formulated now, before the passions of acquisition are running full blast? If there is to be a new policy, what form should it take?

POLICY IS A PLAN TENDERED, NOT A Q.E.D.

A few plain words should be said at the outset. There is nothing in any set of known facts that dictates policy. Facts do not command anyone to do anything. At best facts merely tell us what may be done or must be done, *if* it is deemed desirable to seek some end or objective.

The choice of an end is an act of human will, ethical either in intention or implications, or both. When an individual chooses an end or goal, he takes a position, asserts a good (or evil). A position taken is a position taken. Its truth, or validity, or merit cannot be *proved*. A policy is a value asserted, not a proposition that can be demonstrated in itself. It is good only to those who accept it as good.

THE POLICY PROPOSED

In the matter of the issue before us, the proposition here advanced is simply this: "The government of the United States should take measures calculated to prevent our being drawn or driven into any war in Europe or the Orient."

Respecting this proposition, many positions may be taken by people of intelligence and virtue. It will be useful at this time to review some of these positions.

1. The first position is that the proposition itself is not good, that in certain circumstances good may accrue to the people of the United States and the world from American participation in a war in Europe or the Orient, and that no steps should be taken in advance to prevent it.

2. A second position is that man is an ignorant creature in long-run matters of prognosis, and that the pursuit of immediate interests is as safe a guide as any long-range policy likely to be more or less speculative in nature.

3. A third position is that the proposition rests on the false assumption that mankind can control its own destiny, and has no warrant in our knowledge of the past conduct of mankind.

4. Here the position is taken that the proposition is good and that something can be done by

the government of the United States to accomplish the end posited. Only those who accept this much need consider the paragraphs that follow.

And what is anyone doing who is advancing and pressing this proposition? Horrible as the thought may be for simple minds, it is a fact that such a policy, indeed every large public policy, is *an interpretation of all history*—past, in the making, and to be made. A policy is something deemed desirable and it is pressed (presumably) in the belief that it may be realized in law, and effected in the arrangements of human life. Those who sponsor it make an interpretation of the past facts on which it rests; they assume that it is *possible* to translate the policy into law; and amid the bewildering variables of human affairs they "calculate" that the ends desired *can* and *will* be attained. The path of history is strewn with the wrecks of human policies, designs, aspirations and resolves.

MEANS TO BE CHOSEN

But those who agree that the government of

the United States should seek to avoid war differ fundamentally about the means to be chosen. Those who are dogmatic about the business seem to be sure as Almighty God that they *know* the means to be *chosen* to accomplish the end; and they are equally sure that their opponents in the matter of means are all wrong.

This spirit is admirably expressed by Professor Frederick L. Schuman (*The Nation,* February 12, 1936) when he says that the pending neutrality legislation "is one part lunacy, one part stupidity and one part criminal ignorance of diplomatic and economic realities."

Those who are not dogmatic about the business recognize that they are dealing with probabilities, not certainties. They do not call their opponents on the issue of means lunatics, stupids and criminal ignoramuses. They invite interested parties to remember that they are engaged in the hazardous business of interpreting history and trying to calculate the probabilities of the unknown future. That is a delicate business for any

112

person who, as Mr. Justice Holmes once re-marked, does not imagine himself to be God.

THREE MEANS CONSIDERED

Three means are proposed for keeping the United States out of war in Europe and the Orient. They deserve consideration.

Neutral Rights Shot to Pieces by War. The first involves three lines of action: assert neutral rights, insist upon them, and preserve neutrality. This seems to be the position taken by Professor John Bassett Moore, Senator Hiram Johnson and other prominent leaders in affairs. Incidentally, but not to their discredit, munitions makers, bankers and other pushers of foreign trade support this conception of policy to be pursued.

My objections to this proposal are three in number. The technology of warfare as displayed in the World War and elaborated since that war has shot historic neutral rights into a thousand pieces. If anyone doubts that proposition, let him

113

try to make a list of contraband goods and non-contraband goods in the light of practice between 1914 and 1918. To insist upon the maintenance of those historic rights without efforts to enforce them would be an empty gesture. To enforce them would call for war on the part of the government of the United States. In the third place it was insistence on the right to sell and lend to belligerents and neutrals (engaged in selling to belligerents) that entangled American economic interests in the fate of the World War and brought the country to the necessity of accepting a domestic crash or entering the War. According to my interpretation of history, then, insistence on neutral rights increases the probabilities of entanglement in any war that breaks out in Europe or Asia.

League of Nations Dominated by Imperialist Powers. A second proposal for keeping out of war comes from those who believe in the League of Nations or at least American coöperation with the League powers in efforts to keep peace. They hold that it is impossible for the United States to stay out of war if it comes. They have an interpretation of history to support their belief. At the outset representatives of this school opposed legislation laying an embargo on credits, loans and sales to belligerents and neutrals. Finding this opposition futile, they then concentrated their forces on preventing the mandatory features of the embargo. They would enlarge the power of the President and State Department by giving them discretionary power to marshal the economic forces of the United States "on the side of peace"; that is, in effect, on the side of the League powers or the managing powers of the League in all efforts to maintain the *status quo* in Europe, Africa and Asia against disturbers.

One view, stated baldly, holds that the United States, in the interest of democracy and humanity, should sustain Great Britain, France and Russia against the fascism and despotism of Germany, Italy and Japan. It is a case, once more, of "saving civilization," as some of us thought and honestly believed in 1914.

My objections to "leaving it to the President and State Department" to throw the weight of the country on one side or the other in every pending quarrel in Europe rest upon my interpretation of history. It may be that my ignorance of "diplomatic and economic realities" is abysmal, but I hope that it is not "criminal," as Professor Schuman would have it. At least I have given some attention to the subject.

If I read correctly available diplomatic documents and papers, among the weighty influences behind the World War was the rivalry of the imperialist powers over trade and territory. The several Great Powers were pursuing that line of "national interest." The Secret Treaties and the Treaty of Versailles seem to me to reveal their fundamental motives in preparing for the World War and in carrying it on. I do not say that all can be reduced to economic motives. That would be oversimplification. But I do hold that the imperialist rivalry for trade and territory was among the efficient motives in pre-war diplomacy and in the War itself.

116

Now I may be displaying "lunacy" and "stupidity," to use Professor Schuman's formulas, but it does not seem to me that the League of Nations has changed the old imperialist rivalry in any fundamental respect. The Great Powers of Europe and Asia, whether in or out of the League, are pursuing their "national interests" in the good old way behind the scene, and are employing League methods when convenient. The only exception is Russia, and I do not pretend to know the intentions of the Russian government. It seems inclined to stay at home and attend to its own business of socialism in one country, but appearances may be deceptive.

Now it is my thought that the people of the United States have been badly burned trying to right historic wrongs in Europe, to promote democracy there and to act as mediator in European quarrels. And it is my theory of probabilities that we shall be badly burned again if we keep on insisting that it is our obligation to do good in Europe. That assumes that we know "good" when we see it, and that we can make it prevail in Great Britain, France, Germany, Italy and elsewhere. I am not sure of our ability to know it when we see it and to make it prevail. Yet I am not a foe of those who want to do good nor of good itself. My trouble lies in the fact that greed, lust and ambition in Europe and Asia do not seem to be confined to Italy, Germany and Japan; nor does good seem to be monopolized by Great Britain, France and Russia.

Nor do I trust the omniscience of the President and the State Department. Probably they would not exactly claim it in a pinch. The grounds for my distrust lie in many things we know now about the doings of John Hay, Theodore Roosevelt, Woodrow Wilson, Robert Lansing, E. M. House and others once engaged in exercising discretionary authority behind closed doors. One has only to examine Theodore Roosevelt's private treaty with Japan, the ease with which John Hay swallowed the conception of "the Open Door" formulated in British interests, and the record of Wilson and House on the Secret Treaties to ac-

118

quire at least a faint suspicion of presidential and diplomatic omniscience.

A great deal depends also on the way in which one views American economy and opportunity. In my opinion the United States has the setting for a continental power. For a time Admiral Mahan, Theodore Roosevelt, Henry Cabot Lodge and Albert J. Beveridge were bent on making us an imperialist sea power. They plunged the government into the competition of European and Asiatic imperialisms. They thought that they had "grown up" and ought to make America a big nation, in the British sense—shoving and pushing trade, sticking American noses into every imperialist quarrel and rattling the symbols of American "greatness" everywhere. To business men and farmers they offered "outlets for their surpluses," and for a time they seemed to be on the way to glory. But now the corroding acid of doubt and skepticism has eaten away the old confidence in finding "outlets for surpluses" by the pursuit of that policy. There are many people in the United States who think with General Smedley Butler that imperialism of the Hay-Roosevelt-Mahan type is just a "plain capitalist racket," which the army and navy of the United States ought not to serve at all.

It is my desire to see that "racket" abandoned. I do not think it "pays." It has not worked; for evidence look at the plight and surpluses of industry and agriculture, forty years after the game was started. It is not in keeping with my conception of national grandeur, welfare and virtue.

Tilling Our Own Garden. Having rejected the imperialist "racket" and entertaining doubts about our ability to make peace and goodness prevail in Europe and Asia, I think that we should concentrate our attention on tilling our own garden. It is a big garden and a good garden, though horribly managed and trampled by our greedy folly. Tilling it properly doubtless involves many drastic changes in capitalism as historically practised. Well, with all due respect to the enterprise and virtues of capitalism, I never regarded that "system" as sacred, unchanging and unchangeable. I should certainly prefer any changes that may be required in it to the frightful prospects of American participation in a war in Europe or Asia.

Although it seems banal to some learned economists, I am appalled by the sight of slums, unemployed millions, poverty and degradation on the one side and the immense productive potentials of agriculture and industry in America. I believe it to be the supreme duty of American intelligence to devise ways and means for using most of our "surpluses" at home. It seems to me a good thing to do in itself. I do not see how these alleged "surpluses" can be sold abroad in war or peace. I hear a lot of talk about "lowering trade barriers," but I have seen no demonstration that it can be done or would work if done. Hence I would effect fundamental changes in domestic economy that would diminish the frightful pressure for selling goods abroad to belligerents and neutrals.

If it is a matter of greed, then I prefer following the long-term conception rather than the

121

short-term conception of greed. But the issue is larger than any mere greedy theory of life. It seems to me that a nation engaged in tilling its own garden and refraining from shoving its selling agents and navy into the faces of possible buyers everywhere might set a noteworthy example to mankind. That may be "lunacy" and "stupidity" too, but it is respectfully submitted for consideration by those to whom it appeals.

NEUTRALITY LEGISLATION MAY HELP EDUCATE THE NATION

In view of the foregoing, I support mandatory neutrality—an embargo on the sale of munitions and the extension of credits to belligerents, and a restriction of sales to neutrals engaged in re-selling to belligerents. This involves an amplification of the neutrality legislation passed in February, 1936. In the light of our experience in 1914-1918, we should be giving away most of the stuff sold to belligerents, and (homicidal as it seems to my great and good master John Bassett Moore) I prefer to give the goods to hungry

122

Americans rather than to fighting Europeans. I should put other limitations upon intercourse with belligerents and neutrals in war zones. I should make them all mandatory.

They will be difficult to enforce—at least some of them. But perhaps enough of them can be so enforced as to prevent the bankers and politicians from guiding the nation into calamity as in 1914–1917. At all events, there would not be the opportunity for the kind of backstairs dealing and manipulating that we had in those fateful years.

Most important of all, in my opinion, would be the educative effect of such a statute. The country has learned something about the lines of policy that lead to war. Thousands of citizens are aware of the way in which bankers put up to Wilson the ultimatum of credits or crisis, loans or crisis in 1914–1915. The mandatory provisions of a statute would be perpetual reminders of history. The background of the economic game and greed involved would be etched into law. In the future bankers would have to go to Congress more or less openly, instead of slipping around to the

State Department or the Metropolitan Club to deliver their ultimata and hear what they are "entitled" to hear. To my simple democratic mind this seems a better way.

But all action by the government of the United States would not thus be stopped. It might so happen that participation by the United States in the next or following war would be desirable "in the national interest" or for some great good. If so, the case could be discussed openly and decided openly on its merits by the Congress of the United States, as advised by the President and State Department openly. If we go to war, let us go to war for some grand national and human advantage openly discussed and deliberately arrived at, and not to bail out farmers, bankers and capitalists or to save politicians from the pain of dealing with a domestic crisis.

With awareness of the uncertainty, tragedy and sadness of things, respectfully submitted.

70
71
72

71
78
76
77
79
8
83
85

88